| | | DATE DUE | | |
|---|---|---|---|---|
| | | | | |
| | | | | |
| | | | | |
| | | | | |
| | | | | |
| | | | | |
| | | | | |
| | | | | |
| | | | | |
| | | | | |

# SCIENCE ON THE EDGE

# GENE THERAPY

○ ○ ○ ○

WRITTEN BY
## LINDA GEORGE

BLACKBIRCH®
PRESS

THOMSON

———— ✦ ————™

GALE

San Diego • Detroit • New York • San Francisco • Cleveland • New Haven, Conn. • Waterville, Maine • London • Munich

*For more information, contact*
The Gale Group, Inc.
27500 Drake Rd.
Farmington Hills, MI 48331-3535
Or you can visit our Internet site at http://www.gale.com

Photo credits: cover, pages 4, 9, 12, 14, 15, 16, 18, 27, 28, 31, 33, 34, 36, 38, 41 © CORBIS; pages 5, 6, 13, 24, 37, 40 © PhotoDisc; pages 7 © Blackbirch Press Archives; page 8 © Art Resource; pages 10, 30 © National Institutes of Health; pages 17, 21, 22, 25, 29, 42, 43 © PhotoResearchers; page 19 © The Nobel Foundation; page 32 © Agricultural Research Service, USDA

### LIBRARY OF CONGRESS CATALOGING-IN-PUBLICATION DATA

George, Linda.
  Gene therapy / by Linda George.
    p. cm. — (Science on the edge series)
Includes index.
Summary: Discusses the history of genetic research that led to development of gene therapy and possibilities for future uses.
  ISBN 1-56711-786-4 (hardback : alk. paper)
  1. Gene therapy—Juvenile literature. 2. Genetics—Juvenile literature. [1. Gene therapy. 2. Genetics.] I. Title. II. Series.

RB155.8 .G476 2003
616'.042—dc21                                          2002012530

Printed in the United States of America
10 9 8 7 6 5 4 3 2

Gene therapy is a way to change a person's genes in order to cure defects or disease. Genes—the basic units of heredity—are located on strands of DNA, which is the "blueprint" for the traits, or characteristics, of living organisms, such as a plant or animal. When genes are missing or flawed, an organism suffers from a disease or defect. Gene therapy adds, replaces, or removes genes from DNA so that such diseases can be treated, cured, or even prevented. This process has developed over more than a century of research and experimentation. The knowledge scientists have gained have the potential to lead to a healthier, more productive future for all people.

**Gene therapy can cure diseases by altering the genes that make up DNA.**

This colorful model shows how strands of **DNA** are wrapped together in a "double-helix."

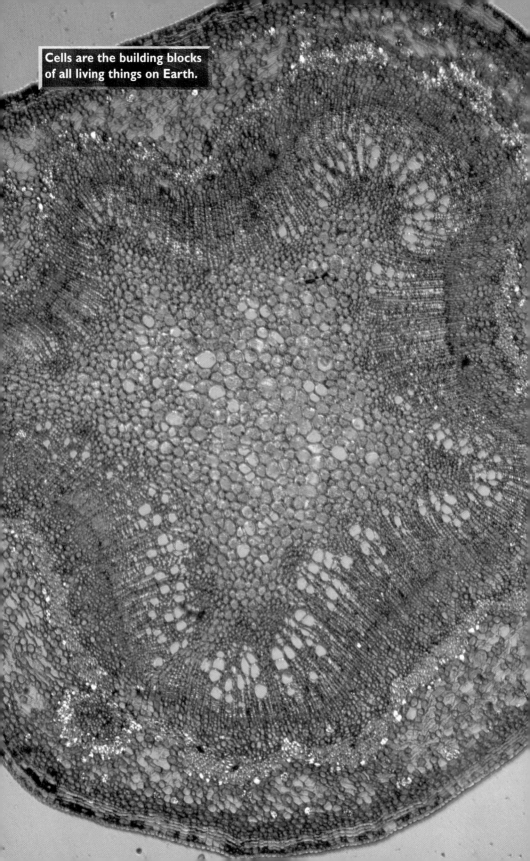

Cells are the building blocks of all living things on Earth.

# CHAPTER 1

# HISTORY OF GENE THERAPY

Human beings tried for centuries to find out how specific physical traits are inherited. Gradually, they realized that traits were somehow related to information found in the cells that make up all living things. A human body is composed of about 100 trillion cells.

Every cell has a center called the nucleus, which acts as the cell's brain. Inside every nucleus are strands of genes called chromosomes. Genes are made of deoxyribonucleic acid, or DNA, which is responsible for the kinds of physical characteristics and

**Chromosomes (shown here) are found in the nuclei of cells. Chromosomes contain thousands of genes which determine physical traits in an organism.**

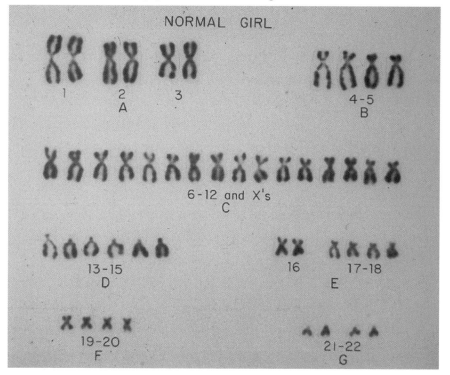

This ancient Egyptian painting shows cattle being brought for inspection before the pharaoh. The Egyptians knew that breeding certain animals would produce more desirable characteristics.

behavior that living things display. Chromosomes contain thousands of genes. A human being has between 30,000 and 100,000 genes that control traits such as eye or hair color, height, and weight. The genes also determine body processes, such as how the cells absorb oxygen and whether the heart beats properly.

Sometimes, chromosomes lack certain genes, or those genes may not work the way they should. This may cause a defect or disease. It took many years to learn how to change the genes to prevent these problems or to help people who suffer from them. The research began with plants and animals.

As early as 3000 B.C., people in Egypt observed animals and their young, such as cattle and their calves. Some cattle provided a lot of meat for food—a desirable characteristic. Others did not—an undesirable characteristic. People at that time did not know about genes and DNA. They only knew that certain animals, when paired for breeding, produced offspring that had the characteristics people wanted. Sometimes, the calves grew up to be the same size and color as their parents. At other times, however, that did not happen. When humans put different cattle with different physical traits together as parents, the appearance of the offspring often changed. People soon learned that they could get the traits they wanted in the offspring if they carefully selected the parents based on their characteristics. This process is called selective breeding.

## GREGOR MENDEL, THE FATHER OF GENETICS

Since animals could be changed through selective breeding, people wondered if plants could be, too. Gregor Mendel, a monk from Austria, did significant work with plant genetics in the mid-1800s. He grew peas in the garden of the monastery where he lived. Some of the pea plants were short and others were tall. Some of the peas were smooth, others wrinkled. As he experimented, Mendel

Austrian monk Gregor Mendel used peas from his garden to establish the basic principles of heredity.

# PASS IT ON

Scientists can read printouts of human DNA strings.

Genetics is the study of how characteristics are passed from parents to offspring. These traits are carried by molecules of DNA found in the egg cells of the mother and the sperm cells of the father.

The human genome (all the genes in a human being) can be compared to a book. The chapters of the book are chromosomes. Each chapter contains paragraphs—the genes. Each paragraph is made up of words that are written in combinations of only four letters—A, T, G, and C. These letters stand for the bases adenine, thiamine, guanine, and cytosine. The human genome book contains a billion words. All the letters, words, paragraphs, and chapters together contain the information needed to construct a human being from a single cell. This information is contained in the nucleus of nearly every cell in the human body.

noticed that, at times, the seed from a tall plant with smooth peas grew to be a short plant with wrinkled peas. He wanted to know how this could happen.

To find out, Mendel carefully chose the characteristics of the parent plants, and then observed the offspring plants. He discovered that some traits were dominant and others were recessive. This meant that when an offspring plant had genes from a tall plant and a short plant, the offspring would be tall because tallness was stronger—a dominant trait—whereas shortness was recessive—a weaker trait.

These discoveries made scientists wonder how these traits were passed from parents to offspring—in plants, animals, and human beings. Mendel's work marked the beginning of the study of genetics, the study of how traits are inherited.

## INSIDE THE NUCLEUS

Decades after Mendel did his work, with the aid of microscopes, scientists observed the structure of cells and the object in the middle of cells—the nucleus. By then, they knew that all living things were made up of cells. They also knew that all life-forms began as a single cell. If one cell was able to become trillions of cells, scientists realized, it had to divide. Microscopes made it possible to see this division occur.

When they watched a cell split into two cells, scientists saw something else inside the nucleus—strands that they had not been able to see before. These strands were chromosomes. Usually, the chromosomes were too tiny and thin to be seen, even with a microscope. Just before the cell divided, though, the chromosomes got thicker and shorter, which made it easier to see them under a microscope.

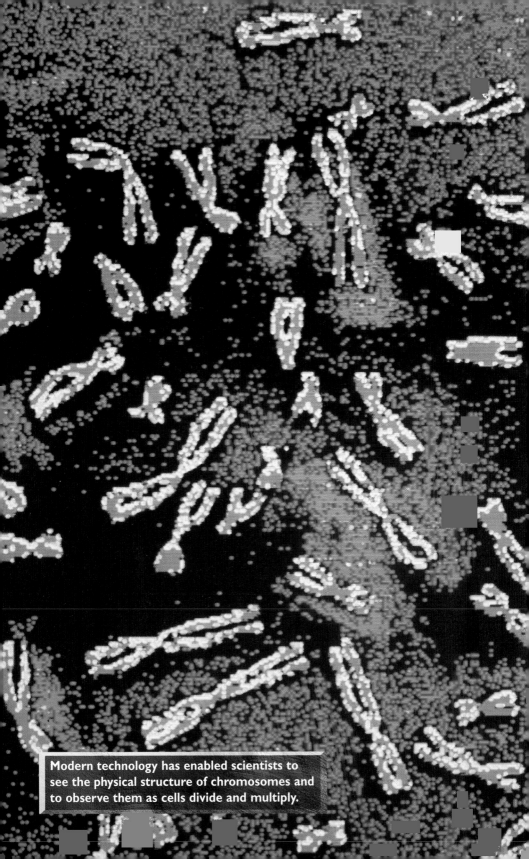

Modern technology has enabled scientists to see the physical structure of chromosomes and to observe them as cells divide and multiply.

After the cell had divided, the two new cells were identical to the original one. This meant that the chromosomes in the nucleus of the new cells were identical to the chromosomes in the original, or mother, cell. Chromosomes contained tiny bits of information that Mendel had called genes. Scientists believed that an individual thing's genes might be a combination of the parents' genes, but they were not sure yet. It would seem to be the case, since the offspring generally looked a little like the mother and a little like the father. Also, if one of the parents was changed, there were often different traits in the offspring, because the new parent's genes were different.

## SELECTIVE BREEDING IN HUMANS?

Until the 20th century, no one had seriously considered the use of selective breeding in humans, but they had definitely thought about it. What if someone were to choose parents according to their characteristics in order to produce a child with specific traits? This had been accomplished successfully in plants and animals.

**Every offspring contains a combination of genetic material from both mother and father.**

Humans were much more complex, though, so scientists were not sure whether it could be done. If it could, then people would have to deal with the issue of whether it would be morally right to control the traits of human beings. Some argued that, if diseases could be eliminated or cured, it would be a good idea to make sure certain human traits were passed on while others were not. Before these decisions could be made, however, scientists first had to learn how to select, control, or change human genes.

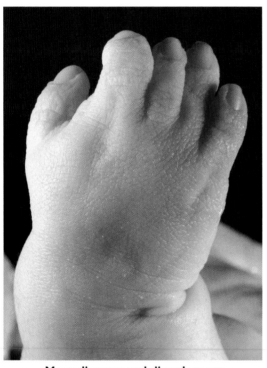

Many diseases and disorders are hereditary—passed on to children through parents. A condition known as brachydactyly, which causes shortened fingers, is a hereditary disease.

In 1905, W.C. Farrabee, an anthropologist at Yale University, studied a family in which some of the members had extremely short, stubby fingers. Because this trait occurred in one parent and some— but not all—of the children in this family, Farrabee concluded that the trait had been passed from the parents to the children through their genes. Short fingers seemed to be a recessive trait, one that was carried by both parents. The trait showed up only when the child received the short finger gene from both parents. If the child received one gene for normal fingers from one parent and one gene for short fingers from the other parent, the child's fingers would be normal, since that gene was dominant.

# DIVIDE AND MULTIPLY

A mouse embryo at the two-cell stage

Every human being begins life as a single fertilized cell. Inside the nucleus of this cell are chromosomes. Chromosomes consist of DNA, which contains genes. Half of the DNA in the cell comes from a father and half comes from a mother.

The single cell divides into two cells, each of which is identical to the original cell. This type of cell division is called mitosis. Each of the two new cells divides, and then those cells divide as well. The chromosomes also divide, so each new cell ends up with an exact copy of the DNA from the original cell.

Eventually, the cells become different types of cells. There are muscle cells, skin cells, brain cells, and dozens of other kinds. In about nine months, that single original cell will have divided into the trillions of cells that make up a human being.

This was the first time a genetic defect could easily be seen and predicted. In order to develop cures and treatments for more serious ailments, however, scientists would have to pinpoint the genes that caused specific diseases or defects. Only then could they replace the missing or defective genes with healthy genes that would not cause disease. Before this kind of gene therapy would be possible, more had to be known about how genes linked to form DNA and how those links might be broken and reattached.

# THE DOUBLE HELIX

**James Watson displays his double-helix model of the DNA structure.**

In 1953, an amazing discovery was made. James Watson and Francis Crick discovered some of the secrets of DNA. They determined that DNA was shaped like a spiral ladder, or staircase. They called its structure the "double helix" (a helix is a spiral). The rungs of the DNA ladder were composed of four substances called bases—adenine (A), thiamine (T), guanine (G), and cytosine (C)—arranged in pairs.

Adenine always bonded with thiamine, and guanine always bonded with cytosine. The sides of the ladder were composed of sugars and phosphates (compounds that contain phosphorous and nitrogen).

When DNA replicated, or copied, itself, the double helix split apart, or unzipped, right down the middle. This separated the pairs

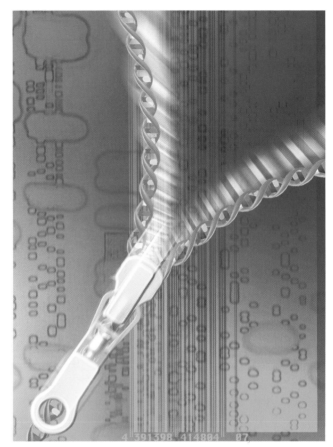

**This computer-generated illustration shows the "unzipping" of the DNA double helix as it replicates.**

of bases. Spare bases, called nucleotides, that usually floated around in the nucleus then bonded with the unzipped bases, to create two spiral staircases that were identical to the original.

Watson and Crick realized that when DNA replicated itself, the pattern of the base pairs was copied exactly. This led them to believe that the order of the base pairs might be what determined a person's genetic traits. They were right. Their work was so significant that Watson and Crick received a Nobel Prize for their discoveries. This information about how DNA was structured and how it duplicated itself was what scientists needed in order to try gene therapy.

# JAMES WATSON AND FRANCIS CRICK

James Watson (second from left)

James Watson was born on April 6, 1928, in Chicago, Illinois. He received his bachelor's degree from the University of Chicago in 1947 and his Ph.D. in genetics from Indiana University in 1950. He then studied in Copenhagen, Denmark, before he joined the Cavendish Laboratory in Cambridge, England. There, he met Francis Crick.

Crick was born on June 8, 1916, in Northampton, England. He received a bachelor's degree in physics from the University College of London in 1937. He was on staff in the molecular biology laboratory at Cambridge, where he worked with James Watson on the structure of DNA at the same time that he studied for his doctorate, which he received in 1954.

In 1953, Watson and Crick published a historic document that described the structure of the DNA molecule. To construct their model, they used a special type of X-ray photography and an electron microscope to look at DNA and determine its structure,

Francis Crick

which resembles a spiral staircase or "double helix." Their work formed the basis of future studies of DNA and genetics. Watson and Crick were awarded the Nobel Prize in physiology and medicine in 1962.

In 1985, James Watson became the first director of the Human Genome Project and once again worked with Francis Crick to expand knowledge of DNA. Watson commented, "We used to think that our future was in the stars. Now we know that it is in our genes."

# FIRST ATTEMPTS AT GENE THERAPY

The first attempt to put gene therapy into practice occurred in 1969. Stanfield Rogers, an American scientist, treated two German sisters who suffered from a disease that caused large quantities of ammonia, a chemical composed of nitrogen and hydrogen, to build up in their bodies. In high amounts, ammonia is poisonous to the body. In the case of the German sisters, it caused the girls to be mentally retarded and epileptic.

Rogers injected a substance into the girls' bloodstreams that he hoped would combine with their DNA to lower the quantity of ammonia in their bodies. The amount he injected turned out to be too small to produce a noticeable effect. Even so, when other scientists found out what Rogers had done, they condemned him for his attempt to perform such a treatment without enough knowledge of the potential effects.

# CUTTING AND PASTING

Scientists knew they would have to locate certain genes on chromosomes before those genes could be replaced or removed. Once the genes were found, a space would have to be made for the new genetic material. The strand of DNA would have to be cut at the specific place where the targeted genes were located. Then, after the new genes had been inserted, the strand would have to be stuck back together again.

In 1972, Paul Berg, a scientist at Stanford University in California, used special substances called restriction enzymes to cut strands of DNA. These enzymes were able to cut the DNA only at points where certain patterns of bases were found. Once a cut was made, Berg inserted new DNA taken from a donor. Then he stuck the ends back together with a different kind of enzyme

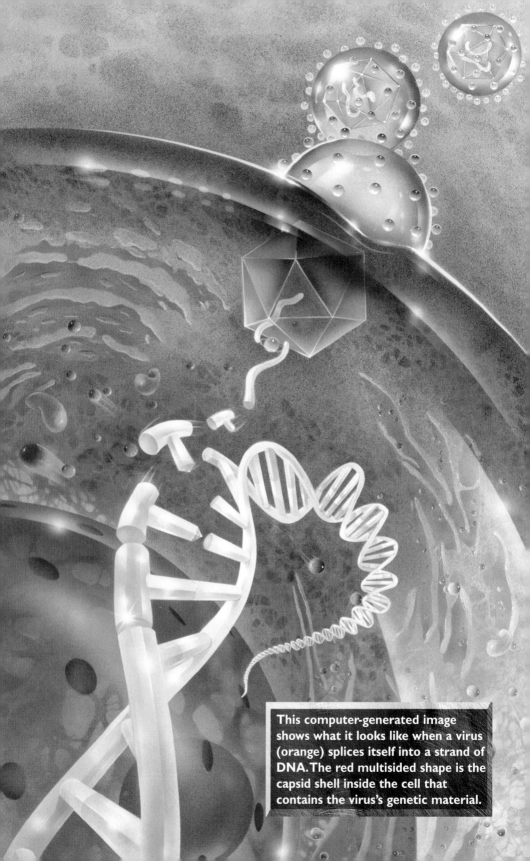

This computer-generated image shows what it looks like when a virus (orange) splices itself into a strand of DNA. The red multisided shape is the capsid shell inside the cell that contains the virus's genetic material.

called a ligase. This process was called gene splicing. (To splice means to stick two ends together.) The ability to splice new genes into a strand of DNA made gene therapy possible. Like Watson and Crick, Paul Berg received the Nobel Prize for his accomplishment. The next step would be to introduce altered genes into a patient's DNA.

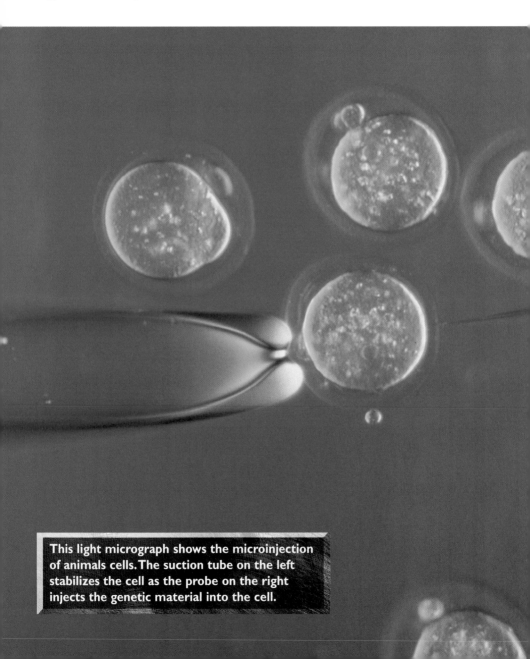

This light micrograph shows the microinjection of animals cells. The suction tube on the left stabilizes the cell as the probe on the right injects the genetic material into the cell.

# MICROINJECTION

In 1976, William French Anderson, who worked at the National Heart, Lung, and Blood Institute in Bethesda, Maryland, came up with a way to inject altered genes into the nucleus of cells with a tiny needle. This technique was called microinjection. By 1979, Anderson had successfully injected a gene for an enzyme into a cell

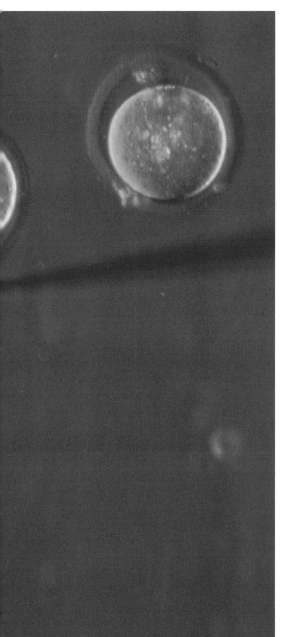

that lacked that gene and, as a result, could not produce the enzyme. The new genetic material he introduced was able to make the enzyme that the original cell could not.

Also in 1979, a researcher named Richard Mulligan, who worked at Stanford University, transferred globin (a kind of protein) genes from a rabbit into the kidney cells of a monkey. After the transfer, the kidney cells began to produce globin. Mulligan's experiment was the first successful transfer of DNA from one mammal to another. After the success of Anderson's and Mulligan's experiments, scientists had the tools they needed to attempt gene therapy on humans.

# WILLIAM FRENCH ANDERSON

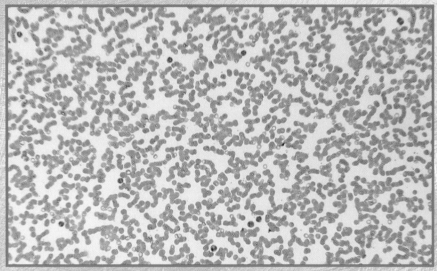

Human blood cells

William French Anderson was born on December 31, 1936.
After he took an undergraduate course in genetics under
James Watson at Harvard University, Anderson delved into
the field. He earned his degree in 1958, then studied with
Francis Crick at Cambridge University. Anderson received
his master's degree in 1960, then returned to Harvard,
where he earned his MD in 1963.

All through his studies, Anderson wanted to learn how to
cure genetic diseases. He studied inherited blood
disorders and tested various ways to introduce new genes
into the human body. Anderson achieved the first
successful gene therapy in 1990 when he treated a so-
called bubble baby for ADA deficiency. In 1991, Anderson
founded a private company called Genetic Therapy. He
has been called "the chief pioneer, impresario, and
champion" of gene therapy.

# CHAPTER 2

## GENE THERAPY TODAY

After many years of research and experimentation, scientists had learned enough about genes and DNA to try gene therapy on a human being. In 1989, gene therapy was attempted with the hope that an existing disease could be cured. Maurice Kuntz, a fifty-two year old man, was terminally ill with cancer. He was selected for a treatment that doctors hoped would destroy the malignant tumors in his body.

This colored electron micrograph shows a killer T-lymphocyte (orange) causing a cancer cell (pink) to undergo self-destruction. T-lymphocytes are part of the body's immune system and are programmed to seek out and attack diseased cells.

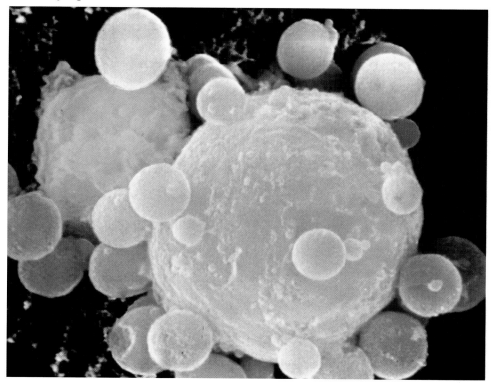

Anderson took white blood cells from one of Kuntz's tumors and inserted new DNA into the cells. Then he duplicated them in the laboratory. In May 1989, 200 million of these genetically altered, cancer-fighting white blood cells—called TILs—were injected into Kuntz's bloodstream. Anderson hoped that the TILs would circulate through Kuntz's body, gather in his tumors, and destroy them. At first, there was no indication that the treatment was working, but gradually, Anderson began to see signs that the TILs were indeed attacking the tumors.

Before the treatment, Kuntz had been expected to live only another three months. Eight months after the treatment, though, he was still alive. The therapy seemed to have worked, at least to some extent. Even so, the cancer did continue to spread throughout Kuntz's body. Kuntz died on April 15, 1990, almost a year after the historic treatment. Although this first attempt had ultimately failed, complete success with gene therapy was not far away.

## FIRST SUCCESSFUL ATTEMPT

The first disease to be treated effectively with gene therapy was severe combined immunodeficiency, or SCID. A child born with SCID does not have the gene needed to produce adenosine deaminase (ADA), a substance that enables the body to fight infection and illness. Before 1990, babies born without the ADA gene had no immunity to disease. To survive, they had to live in a completely sterile environment, nicknamed a "bubble." If a "bubble baby" left the germ-free enclosure, he or she risked death from even the slightest infection.

A treatment for the disease—a drug called PEG-ADA—provided the infection-fighting enzyme that the body lacked. There were some big problems, however. PEG-ADA had to be injected weekly,

The child in this photograph needs to be kept in a completely sterile environment because he suffers from a severe hereditary immune deficiency.

without fail, and it cost about $200,000 per year. The injections, given in the hip, lasted 20 seconds and were terribly painful.

Scientists decided to try to replace the missing gene in the cells of a child with ADA deficiency. They would inject new genetic material that contained the missing gene. If the treatment worked, the weekly injections of PEG-ADA might no longer be necessary. They decided to continue the PEG-ADA injections during the gene therapy procedure, though, just to be sure that the child was protected from infection.

Ashanthi (Ashi) DeSilva, a three-year-old from Ohio, was the first child to receive this therapy. On September 14, 1990, Anderson conducted the first of eight largely successful, historic treatments that continued for ten months. About one-fourth of Ashi's immune system cells were repaired. As a result, for the first time in her life, Ashi was able to swim in a pool and play with other children without fear that she might contract an infection and die. When she did come down with an illness, such as a cold, it cleared up in the same way it would in a normal child.

Ten years later, in 2000, two scientists from France, Marina Cavazzana-Calvo and Alain Fischer, improved the gene therapy treatment and cured two babies who had an especially severe form of SCID. They took bone marrow from the babies and removed the stem cells from it. Then, they mixed the stem cells with a virus whose DNA had been altered to carry the gene needed to produce immune system cells, and injected the mixture into the bloodstream. Within two weeks, it became obvious that the babies' immune systems had begun to work. After a few months, the babies were cured. They were able to live at home without fear of infections that would have killed them before the treatment. Several years later, the children were still completely healthy.

Doctors Marina Cavazzana-Calvo and Alain Fischer pioneered major advancements in gene therapy to help with immune disorders.

# STEM CELLS

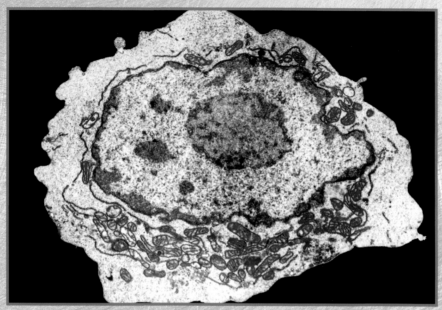

Colored electron micrograph of a stem cell

Stem cells are the cells in an embryo that develop into the different types of cells that make up the human body. Stem cells have been found not only in embryos but also in bone marrow, skin, muscles, and the brain. If stem cells from one part of the body are injected into another part of the body, they become cells of the new part and begin to grow there.

Scientists hope to be able to obtain stem cells from all possible sources in the body for use in laboratory experiments. Stem cells could be one of the most beneficial tools for gene therapy in the future. They might eventually be used to grow new organs or tissues, including livers, hearts, and even parts of the brain and spinal cord, to be used as replacements for damaged or defective human organs.

# OTHER SUCCESSFUL TREATMENTS

Since the 1980s, scientists have studied many diseases to learn how they might be treated with gene therapy. Among these diseases are sickle-cell anemia, hemophilia, and certain types of muscular dystrophy. Also targeted are diabetes, cancer, and heart disease. For some of these diseases, drug therapy is currently used to control symptoms and improve the quality of the patient's life. Altered or new genetic material, however, might eventually make drug therapy unnecessary.

One form of gene therapy in wide use today is in vitro fertilization. (In vitro means that fertilization takes place in a laboratory rather than inside the body.) Couples who have been unable to have children sometimes choose this procedure to achieve pregnancy. The mother's eggs are fertilized by the father's sperm in a laboratory. Once the egg begins to divide, it becomes an embryo. A doctor then implants the embryo into the mother's uterus, where

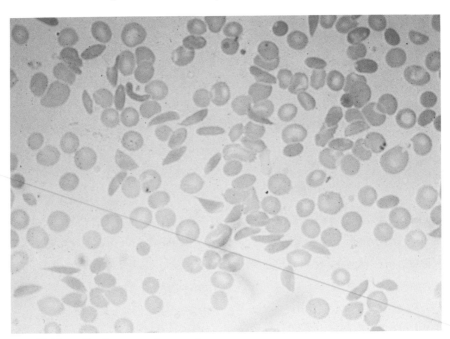

Sickle cells shown under a micrscope

it develops just as it would under normal circumstances.

Before the implantation, several fertilized eggs are examined for their genetic makeup. Diseases such as Tay-Sachs and fragile X syndrome (a condition that causes mental retardation in males) can be detected, and embryos that have these abnormalities are discarded. Only an embryo that is free of defects is implanted in the mother.

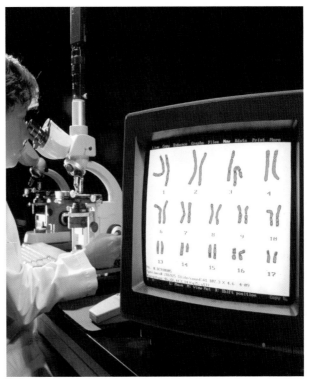

**Genetic screening enables doctors to predict or identify certain hereditary problems before an offspring is conceived or born.**

This procedure not only helps reduce the risk of birth defects, but it also allows the parents to choose the baby's sex. A female has two X chromosomes, whereas a male has one X chromosome and one Y chromosome. An embryo with the desired combination of chromosomes can be selected in order to ensure that the baby is whichever sex the parents desire. This is possible because doctors know how to find and identify the X and Y chromosomes. If scientists knew where all the genes were located on all of the chromosomes (humans have forty-six chromosomes), they would be able to develop procedures to treat, cure, or prevent virtually all defects and diseases through gene therapy. The first step, though, requires a full knowledge of the genetic code contained in human DNA.

# GENETICALLY ALTERED PLANTS AND ANIMALS

Tomatoes

Gene therapy can also help people by making improvements in the foods they eat. Since the mid-1990s, scientists have attempted to change the genetic makeup of foods to make them more healthful for people to eat. People across the United States today eat many genetically altered foods. Soybeans, corn, and canola are just three genetically improved items that are used in processed foods such as pizza, cookies, and salad dressing.

In 13 countries around the world, 50 kinds of crops on 130 million acres have been altered genetically. Tomatoes, rice, apples, and squash are a few of the foods that have been improved through the insertion of new genes into their DNA. About 100 more crops are also being tested to see if they can be made better.

To make these improvements, scientists combine the genes of different species. Plants with new genes are called transgenic plants. Transgenic means that genes move from one species to another. The genes do not have to come from two animals or two plants. A gene from a bacterium found in soil may be inserted into the genes of corn plants. This enables the corn plant to produce a natural insecticide to kill corn borer caterpillars before they can cause damage. If a rat gene is inserted into lettuce, the lettuce produces more vitamin C. Genes from a certain type of moth can be inserted into apple trees to make their leaves, stems, and fruit resistant to harmful bacteria. Genes from bees and moths can help potato plants fight a disease called potato blight fungus.

One of the most promising transgenic foods is golden rice. This new rice is created when two genes from a daffodil and one gene from a bacterium are inserted into regular rice. These combined genes produce beta-carotene, which is not normally present in rice.

Beta-carotene produces vitamin A. Golden rice could provide vitamin A to some of the 100 million to 140 million children in the world who suffer from vitamin A deficiency, which can cause blindness or death.

Golden rice field

In addition to improved, genetically altered plants, there are also transgenic animals. Atlantic salmon have been treated with hormone genes to make them grow faster, and chickens can be treated with genes to make them less aggressive and easier to handle.

Although there are many potential benefits from genetically enhanced foods, some people do not want to eat them. They worry that if they eat foods whose genes have been altered, their own genes might change, too. Many tests are now under way to see if such alterations could actually take place. Scientists are eager to determine that the process is safe. If they can be sure that genetically altered plants and animals are absolutely safe to eat, it could mean the end of world hunger and better nutrition for all of the people on Earth.

Pig

# HUMAN GENOME PROJECT

The Human Genome Project (HGP) began in 1985. (The total of all the genes in a human being is called the human genome.) The project's goal was to map all the base pairs (adenine paired with thiamine, and guanine paired with cytosine) in human DNA so they could be deciphered and identified as specific genes.

The first director of the project was James Watson, one of the scientists who had discovered the structure of DNA in 1953. Watson worked with his former research partner, Francis Crick,

**Dr. Craig Venter (left) hands Senator Pete Domenici a copy of the map that shows the sequencing of the human genome during a special ceremony in Washington, D.C., February 12, 2001.**

THE SEQUENCE OF THE HUMAN GENOME

as well as other scientists and researchers from around the world. Their goal was to map the human genome within twenty years. By 1997, only about half of the genome had been sequenced.

In 1997, J. Craig Venter and others from his company, Celera Genomics Corporation, came up with a way to map the genome faster and announced that they were ready to try to finish the mapping before the HGP did. The race was on!  In the end, the two research teams finished at about the same time.  On June 26, 2000, President Bill Clinton congratulated both groups when he announced the completion of the first draft of the human genome.

The next step was to figure out how the base pairs were grouped into genes, and what they controlled in the human body. In 2001, hundreds of geneticists—scientists who study genes—worked together.  They shared the work they had done on various parts of the genome to fill in as many genes as they could.  They also guessed the position and function of unknown genes.  Eventually, they identified between 30,000 and 40,000 genes.  This was only one-third of the number of genes they had previously thought they would find.

These accomplishments laid the foundation for successful gene therapy in the future.  Mapping the genome enabled scientists to locate specific genes that caused disease.  Today, there are treatments and cures for some diseases that were not possible before the human genome was mapped.  More genes are identified every day.  These discoveries will lead to even more treatments and cures.

# CHAPTER 3

## WHAT LIES AHEAD

The list of possibilities for what lies ahead in gene therapy sounds like science fiction. To be able to grow new organs for transplants, to clone beloved pets or children who have died, or to grow teeth, skin, and cartilage in a laboratory—all of these sound more like the plot of a futuristic movie than actual scientific procedures. Yet all these accomplishments will be possible thanks to the knowledge scientists first gained in the 20th century about how the human genome is constructed and how genes control human characteristics and functions.

Perhaps the most important result of all this knowledge will be the extension of the human life span. Within the 21st century, the average life span will increase to at least 100 years. The primary reason will be the elimination of

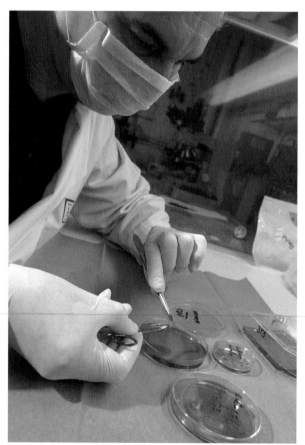

Scientists are seeking ways to use gene therapy techniques in the battle against HIV—the virus that causes AIDS.

**In the near future, the average human life span
is expected to increase to at least 100 years.**

virtually every disease that plagues humans by the end of the
century. Gene therapy and procedures that work along with it will
prevent countless deaths from diseases such as cancer, Alzheimer's,
and diabetes. When gene therapy becomes routinely successful,
people will live much longer and be healthier than ever before. The
more scientists learn about all types of genes, the more treatments
they will find for diseases caused by genetics. The day will come
when all diseases that result from genetic defects can be cured.

The majority of these procedures will depend on what scientists
are able to accomplish through the use of stem cells in gene
therapy. Stem cells are unique cells in the body that are able to

**Certain gene therapy techniques already enable scientists to explore ways of creating powerful and effective vaccines and antibiotics.**

develop into virtually any type of cell the body needs. Stem cells may be used to grow replacement cells for all types of tissue in the body. Artificial flesh has already been created in the laboratory and used with great success on burn victims.

Scientists predict that in the 21st century, gene therapy techniques will develop new antibiotics that will be able to cure viral diseases that cannot now be treated. New vaccines will also be developed. Today, vaccines prevent diseases such as polio and smallpox, as well as measles, mumps, and some types of hepatitis. Within this century, gene therapy will help produce vaccines to prevent the common cold and AIDS, along with other diseases for which there are currently no vaccines available.

Although scientists do not yet know exactly how these amazing things might be done, new techniques are constantly developing and those who work with gene therapy are confident that all of these medical miracles will come into being before very long. For now, even while they search for better ways to use gene therapy, scientists concentrate on using tested methods to find new treatments for common diseases. The disease scientists want to be able to cure most of all is cancer—in part because there is nothing that people can do to avoid developing certain types of the disease.

# A CURE FOR CANCER?

Most current gene therapy trials involve cancer research. Every day, new knowledge is gained and scientists learn how to control and sometimes cure certain types of cancer that were previously fatal. One form of Hodgkin's disease, a type of cancer that causes tumors in the lymphatic system and other parts of the body, can sometimes be cured in part thanks to a gene therapy technique that involves bone marrow treatment. One type of skin cancer, called melanoma, kills about 7,500 people each year. Recently, scientists have discovered a gene that, when damaged, may contribute to melanoma. They hope to develop a way to mend the damaged gene and stop the growth of melanoma cells.

Scientists also try to inject cancer cells with a gene that will make them more susceptible to cancer-fighting drugs. If these experiments are successful, cancer drugs will leave healthy cells alone and attack only cancerous cells. This would solve one of the most difficult problems in cancer treatment—how to fight the cancer without weakening the healthy parts of the body. In time, there will be hundreds of new cancer drugs and treatments, which will prevent thousands of deaths each year, all as a result of what scientists are now learning about how to identify and manipulate genes.

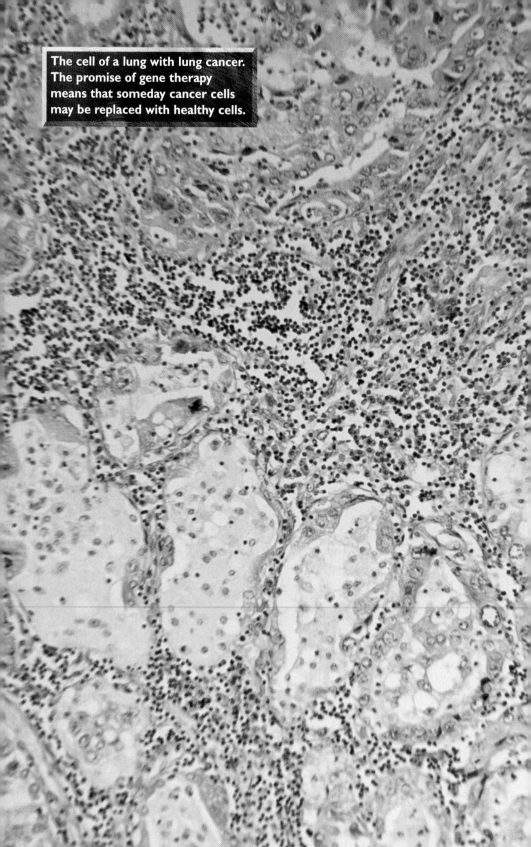

The cell of a lung with lung cancer. The promise of gene therapy means that someday cancer cells may be replaced with healthy cells.

# OTHER BENEFITS

Gene therapy will not only provide new treatments for diseases, but may also prevent or repair physical defects and malformations, such as dwarfism and Down syndrome. Children who are born blind or deaf may be able to see and hear with the help of implants of healthy, genetically altered living cells. Nerve tissue grown from stem cells will be able to reverse spinal cord damage that causes paralysis. Already, doctors have been able to restore partial movement in the arms and legs of paralyzed victims of spinal cord damage. To do this, doctors inject a special type of white blood cell, called a macrophage, into the damaged cord tissue. The macrophages repair some of the damage to the nerve tissue, which allows movement to occur where there was no chance of movement before. Amazing breakthroughs such as these will not come without problems, however.

**This image shows what it looks like when a macrophage creates a new connection—-or synapse—with the axon of a nerve cell.**

# PROBLEMS TO RESOLVE AND DECISIONS TO MAKE

Even after scientists have developed cures for all diseases, everyone will not necessarily have access to those cures. At first, the treatments will cost a great deal of money. Only the wealthy will be able to afford them. Eventually, though, the treatments may cost less, so more people will be able to benefit from them.

All of the gene therapy attempts now being made involve changes to soma, or body, cells. These genetic changes will help the patient who is treated but will not be passed along to his or her offspring. Gene therapy to germ line, or reproductive, cells, on the other hand, would make changes that would be inherited by the offspring of the treated person. Currently, germ line gene therapy is not practiced, even though it is much easier to do than soma cell gene therapy. Scientists have agreed that germ line alterations might produce undesirable side effects that could not be reversed or controlled. It

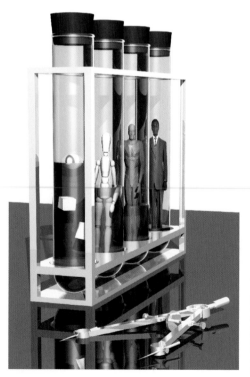

might be possible, for example, to create a different species of human beings—a potentially scary idea!

Scientists agree that, in time, changes will be made to germ line cells when such a change means the difference between life and death or allows for a substantial increase in the quality of life for the individual and his or her future offspring. Which

**The techniques and potential uses of gene therapy are controversial. Many people feel that this technology will lead to "test tube" humans who are genetically programmed to be perfect.**

changes will be allowed have not yet been determined. People of the 21st century will have to answer some difficult questions about what types of gene therapy should or should not be practiced.

A technician loads data into an automated DNA sequencer. Information provided by the full mapping of the human genome will have potentially unlimited use to the scientists and researchers of the future.

## AN EXCITING FUTURE AHEAD

Since the middle of the 19th century, scientists have learned more about the human body and how it works than in all the preceding centuries combined. It is reasonable to believe that the same will be true of the 21st century. In time, the human genome will be fully mapped and labeled from beginning to end, and scientists will know exactly which genes control the thousands of functions and characteristics of the human body. Once this knowledge has been obtained, it will be possible to make changes in the genes to improve quality of life for everyone on Earth. Some of what gene therapy promises to bring about sounds like science fiction. Yet, in time, this science fiction will become fact and will open the way to even more fantastic accomplishments in the future.

# GLOSSARY

**antibiotic** a drug that kills bacteria and is used to cure infections and diseases.

**bases** the four substances found in DNA—adenine, thiamine, guanine, and cytosine.

**chromosomes** the parts of a cell that carry genes.

**cloning** making exact copies of a gene, cell, or organism.

**DNA** deoxyribonucleic acid, the substance genes are made of; DNA is shaped like a spiral ladder, or helix, with pairs of bases as the rungs and sugar and phosphates as the sides of the ladder.

**double helix** another name for DNA.

**egg** a cell within the body of a woman or female animal that, when fertilized, grows into a new individual.

**embryo** a fetus in its earliest stage of development.

**gene splicing** combining genetic information from two or more organisms.

**gene therapy** the treatment of genetic disorders through the insertion of healthy genes into cells to replace defective genes or supply missing genes.

**genes** sections of DNA that determine the characteristics of a species or organism.

**genetics** the study of ways personal characteristics are passed from one generation to another through genes.

**genome** all the information contained in the DNA of a species.

**germ line cells** reproductive cells, such as eggs and sperm.

**implant** an organ or device inserted into the body through surgery.

**in vitro fertilization** fertilization of an egg by a sperm in a laboratory.

**microinjection** the use of a tiny needle to inject genetic material into the nucleus of a cell.

**nucleus** the central part of a cell which contains the chromosomes.

**selective breeding** choosing parents in order to control the characteristics of the offspring.

**soma cells** body cells.

**sperm** one of the reproductive cells from a man or male animal; sperm is able to fertilize eggs in a female.

**stem cells** cells in an embryo that eventually become all the different types of cells in the body.

**transgenic** combining genes from two organisms.

**vaccine** a substance that can be injected or given to a person orally so the person will not get a particular disease.

# FOR FURTHER INFORMATION

*Books*

Boon, Kevin Alexander. *The Human Genome Project: What Does Decoding DNA Mean for Us?* Berkeley Heights, New Jersey: Enslow Publishers, 2002.

Cefrey, Holly. *Cloning and Genetic Engineering.* Danbury, CT: Childrens Press, 2002.

Gonick, Larry, and Mark Wheelis. *The Cartoon Guide to Genetics.* New York: Harper Perennial, 1991.

Graham, Ian. *Genetics: The Study of Heredity.* Milwaukee, WI: Gareth Stevens, 2002.

Hyde, Margaret O., and John F. Setaro. *Medicine's Brave New Worth: Bioengineering and the New Genetics.* (for grades 7–12) Brookfield, CT: Twenty-First Century Books, 2001.

Klare, Roger. *Gregor Mendel: Father of Genetics.* Springfield, NJ: Enslow Publishers, 1997.
*Good descriptions of Mendel's experiments and the traits he studied in pea plants. Excellent glossary.*

Marshall, Elizabeth L. *The Human Genome Project: Cracking the Code Within Us.* Danbury, CT: Franklin Watts, 1996.

Snedden, Robert. *Cell Division and Genetics.* Portsmouth, NH: Heinimann Library, 2002.

Yount, Lisa. *Genetics and Genetic Engineering.* NY: Facts On File. 1997.

*Websites*
Kids Genetics
Understandable information on DNA, genes, and heredity.
http://www.genetics.gsk.com/kids/index_kids.htm

Kids Only
Basics of genetics, gene games, glossary.
http://www.genecrc.org/site/ko/index_ko.htm

Genomics and Its Impact on Medicine and Society
A 2001 Primer
Easy to understand explanations of basic concepts of genetics, the Human
Genome Project, and gene therapy.
http://www.orni.gov/hgmis/publicat/primer2001/index.html

Tiki the Penguin
Tiki's Guide to Genetics
Information on genes, cell division, and genetic engineering.
http://www.oneworld.org/penguin/genetics/home.html

# ABOUT THE AUTHOR

Linda George has been a professional writer for more than twenty years and has written more than three dozen books for children. She lives in the New Mexico mountains with her husband, Charles, who is also a writer.

# INDEX